大鼠害怕香味剂

Perfume for Rats

Gunter Pauli

冈特·鲍利 著

凯瑟琳娜·巴赫 绘
唐继荣 译

丛书编委会

主　任：贾　峰

副主任：何家振　闫世东　郑立明

委　员：牛玲娟　李原原　李曙东　李鹏辉　吴建民
　　　　彭　勇　冯　缨　靳增江

特别感谢以下热心人士对译稿润色工作的支持：

王必斗　王明远　王云斋　徐小怙　梅益凤　田荣义
乔　旭　张跃跃　王　征　厉　云　戴　虹　王　逊
李　璐　张兆旭　叶大伟　于　辉　李　雪　刘彦鑫
刘晋邑　乌　佳　潘　旭　白永喆　朱　廷　刘庭秀
朱　溪　魏辅文　唐亚飞　张海鹏　刘　在　张敬尧
邱俊松　程　超　孙鑫晶　朱　青　赵　锋　胡　玮
丁　蓓　张朝鑫　史　苗　陈来秀　冯　朴　何　明
郭昌奉　王　强　杨永玉　余　刚　姚志彬　兰　兵
廖　莹　张先斌

目录

大鼠害怕香味剂	4
你知道吗?	22
想一想	26
自己动手!	27
学科知识	28
情感智慧	29
艺术	29
思维拓展	30
动手能力	30
故事灵感来自	31

Contents

Perfume for Rats	4
Did you know?	22
Think about it	26
Do it yourself!	27
Academic Knowledge	28
Emotional Intelligence	29
The Arts	29
Systems: Making the Connections	30
Capacity to Implement	30
This fable is inspired by	31

在一栋房子的地下,两只大鼠正在休息,这地方让它们远离了所有危险,很安全。

"我真高兴,人们不再被允许使用那些糟糕透了的毒药,那东西我们会误认为是食物,然后慢慢地杀死我们。"

A couple of rats are taking a rest underneath a house, where they are safe from any danger.
"I am so glad that people are no longer allowed to use that very bad poison – the one that we would mistake for food, and that would slowly kill us."

两只大鼠正在休息……

A couple of rats are taking a rest ...

继续使用诱捕器……

Continue to use traps ...

"好吧。那种吃了能在胃里烧出孔来的可怕东西只是在数百个孩子因此死亡之后,他们才不得不禁止使用它。"

"很多痛恨我们的人继续使用诱捕器。这些诱捕器能打碎我的头骨,或切断我的腿脚,让我在遭受极度痛苦的同时流血而死。"一只大鼠说。

"Well, it was only after hundreds of their children died after eating that horrible stuff, that burns a hole in your stomach, that they were forced to ban it."

"Many people who hate us continue to use traps. Ones that can crush my skull, or chop off my legs, so I will bleed to death, while suffering excruciating pain," one rat says.

"想想看,人们甚至已经发明了大鼠'电刑室'!"

"什么?我以前从来没有看到那些个东西。"

"你过去可能没有留心。它是一个小走廊,末端放有食物。当你进去,得到食物并转身时,你的鼻子和尾巴就触碰到电线接通了电路,然后……'砰'!"

"Imagine, people have even invented electrocution chambers for rats!"

"What? I have never seen those before."

"You probably did not notice it. It is a little tunnel with some food at the end. When you get in, get the food and turn around, your nose and tail touch a wire, and… Bang!"

大鼠"电刑室"!

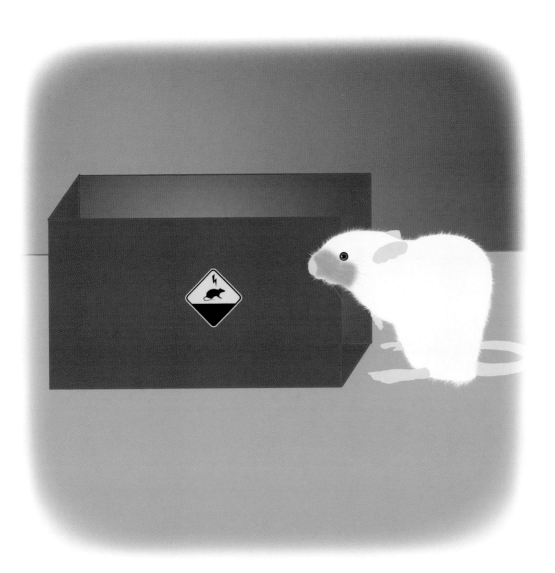

Electrocution chambers for rats!

带着食物径直后退着出来

Walked out backwards with the food

"你怎么对这种东西了解得这样清楚?"

"我曾经非常凑近地看过。因此,我才不会在走廊里转悠,而是径直后退着出来。等到他们的电池没电的时候,我还会带着食物走开。"

"How come you know this so well?"
"I had a close look at one. So instead of turning around in the tunnel, I just walked out backwards, and walked away with the food – while their batteries ran flat."

"人们到什么时候才能意识到,长久以来无论他们怎么千方百计地想杀死我们,我们仍然生活在他们周围,而且数量众多?"

"这是因为,他们一方面试图杀死我们,一方面不断地为我们提供非常好的食物。如果人们不希望我们出现在周围,那最好停止浪费,不再随处留下这么多的食物喂饱我们。"

"When will people realise that despite them having tried for so long to kill us, we are still around – in force?"

"That is because while they are trying to kill us, they continue to feed us so well! If people do not want us around, then they had better stop wasting so much food, leaving it lying around for us to feed on."

随处留下食物喂饱我们……

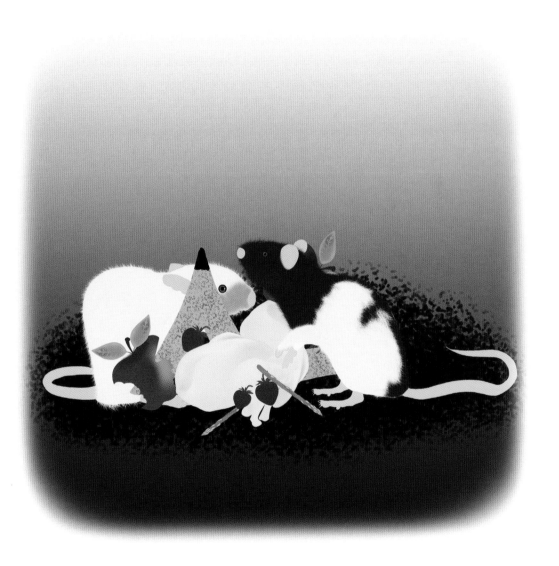

Leaving food around for us to feed on ...

掺了辣椒的食物？

Food mixed with chilli?

"似乎的确有些人变得更聪明了!你有没有因为吃到掺了辣椒的食物,辣得火烧舌头似的?"

"噢,可不是。我看到一只鸟儿吃得有滋有味,就想我也应该来一点儿。结果,辣得我呀,喝多少水都不管用。"

"辣椒真讨厌,不过倒是没人用它对付我们。"

"It does seem that some people are getting smarter. Have you ever burned your tongue eating food mixed with chilli?"

"Oh yes, I saw a bird feasting on it and thought I will get my share. It burnt so badly that no amount of water could bring me any relief."

"Chilli is a real nuisance, yet almost no one uses it."

"我知道为什么!你看,他们要的不是虐待我们,而是杀死我们。前些日子,我吃到了一小块浸泡在肉酱里的软木塞,它完全堵塞了我的肠道。我以为我要死了,结果它终于弹出来……噢,我又活啦!"

"很高兴你活下来,兄弟!不过现在他们又有了一种新的玩意儿:声屏障。"

"声屏障是怎么工作的?"

"I know! You see, they don't want to mistreat us; they want to kill us. Recently, I ate a piece of cork, drenched in meat sauce, which completely blocked my gut. I thought I would die until it finally popped out and ... oh what a relief!"

"Glad you survived, brother! And now they have another new thing: the noise barrier."

"How does that work?"

……一小块软木塞……

... a piece of cork ...

……哔哔的蜂鸣声，让你的耳朵很不舒服……

... beeping sounds that hurt your ears ...

"你有没有听到过那些'哔哔'的蜂鸣声,让你的耳朵很不舒服?人们似乎听不到,但它真的把我吓坏了。我讨厌这个声音,但必须承认,过了不久也就习惯了。"

"但是对我来说,最糟糕的是他们这种在玉米芯上涂抹上草药和油的新型香味剂。最初它被用来驱赶床垫中的臭虫。这是一种天然产品,不会造成直接伤害,但是,孩子,它真的真的太!臭!啦!这种东西会让我逃得远远的,再也不要回去。"

"Have you come across those beeping sounds that hurt your ears, deep inside? It seems people do not hear it but it really scared me. I hated it, but must admit after a while I got used to it."

"But for me the worst is this new perfume they have: herbs and oils on corn cobs. It was first used to chase bedbugs out of mattresses. Its natural, and it does not hurt, but boy, does it stink! It will make me run far away – never to come back."

"好吧!如果我们必须为食物而战斗,就让我们学习狐猴,来个'敌不犯我,我不犯人'。我宁可忍受刺激性气味,也不愿死于毒药、敲头、电击,或者屁股上有个妨碍我上厕所的塞子!"

"你敢再说一遍?!"他的朋友回应道。

……这仅仅是开始!……

"Well, if we have to fight for food, let's learn from the lemurs. I'd rather put up with pungent smells, than be killed by toxins, or knocked over the head, receive an electric shock or have a plug in my butt that prevents me from going to the toilet!"

"You can say that again!" his friend replies.

... AND IT HAS ONLY JUST BEGUN!...

……这仅仅是开始！……

... AND IT HAS ONLY JUST BEGUN! ...

Did You Know?
你知道吗？

Rats do not like the smell of peppermint, daffodils and hyacinths. A cotton ball soaked with peppermint oil can drive rats out of hiding. Pet food and bird feeders attract rats.

大鼠不喜欢薄荷、水仙花和风信子的气味。一团用薄荷油浸泡过的棉花球能将大鼠从藏身之处驱赶出来。宠物饲料和野鸟喂食器则招引大鼠。

Rats and mice are easily frightened by strange and unfamiliar noises. Sounds we as humans cannot hear, may at first scare rats away but they, however, adjust to it and that makes rat control with sound ineffective.

大鼠和小鼠很容易被奇怪或不熟悉的声响惊吓到。一些我们人类不能听到的声音可能最初会让大鼠惊慌失措、四处逃窜，然而它们能调整自身来适应那些声音，这使得用声音进行大鼠防治的效果不好。

The best way to keep rats away is by keeping grains, beans, sugar and other food in sealed glass or metal containers, rather than sacks and bags.

让大鼠远离的最好方式就是用密封的玻璃或金属容器存放谷物、豆类、糖类和其他食物,而不是用大麻袋和布袋。

There are an estimated 10 rats for every resident of New York City. Before rat poison was banned in the USA, 12 000 to 15 000 per year reports were received of children being poisoned by these chemicals.

在纽约市,估计居民和大鼠的平均比例是1∶10。在鼠药被禁用前,美国每年会接到1.2万~1.5万例有关儿童被这些化学物质毒害的报告。

Poison residue has been found in 75% of dead animals studied by the California Department of Fish and Wildlife, from 1995 to 2010. It seems that rat poison was killing more wildlife than it was decimating rat populations.

1995 年—2010 年，加利福尼亚州鱼类和野生动物部研究的死亡动物中，有 75% 发现毒药残留物。看来比起需要消灭的大鼠种群，鼠药在过去杀死更多的是野生动物。

Integrated rat control implies the regeneration of owl and hawk populations, which have been decimated by rat poison. One raptor can hunt and eat six mice and rats in one night.

大鼠综合防治意味着已经被鼠药消灭的猫头鹰和鹰类种群的再生。每只猛禽一晚上能捕食 6 只小鼠或大鼠。

Rats die from eating blue cheese. The fungus (mould) produces chemicals that are likely to lead to rats developing tumours. Peanut butter can clog a rat's throat and even makes it choke.

大鼠吃蓝纹奶酪会致死。真菌（霉菌）产生的化学物质可能会导致大鼠患上肿瘤。花生酱可能堵塞老鼠的喉咙，甚至使它窒息。

The legend of the Pied Piper of Hamelin (Germany) dates back to the Middle Ages. The city hired the Piper, a rat-catcher to lure rats away. When the citizens did not pay, he used the same technique to lead the children away.

德国哈默尔恩城"花衣魔笛手"的传说可以追溯到中世纪。那位穿花衣的风笛手是一位捕鼠行家，这座城市雇佣他来诱使老鼠离开。当市民不肯付钱时，他用同样的方法把孩子们也带走了。

Think About It

想一想

Are you in favour of killing rats with traps, or do you prefer other methods?

为了杀死大鼠,你赞成用诱捕器,还是更喜欢其他方法?

If you caught a rat alive, what would you do with it?

如果你活捉了一只大鼠,你会拿它怎么办?

Are their particular odours that you do not like? If these smells were around, would that make you go away?

你是否不喜欢某些特定的气味?如果你身边有这样的气味,你会离开吗?

Can we learn from Nature, instead of just learning about Nature?

我们能不能向大自然学习,而不仅是了解大自然呢?

Ask around and find out if anyone living close to your home has a problem with rats or mice. If you find a household with rat problems, see how clean it is. Is there dog food left outside? Are there bird feeders? Are the gutters clean and is all waste carefully disposed of? Is food stored in glass or metal containers, or is it easy for vermin to find food in cardboard boxes and plastic packaging? Rats and mice thrive when we do not take good care of our homes and make it easy for them to access any food anytime. Now check where you are living and make sure that vermin does not have any chance of starting a family next to yours. Write a list of simple procedure that can be used to reduce the risk of rats living as your neighbours.

问问周围的人，看看你家附近是否有人遭遇鼠患。如果你找到哪家面临这个问题，就考虑如何清除它。室外还留有狗粮吗？有野鸟喂食器吗？水槽是不是已经清理干净，所有废物是否小心地处理掉了？食物是否储存在玻璃或金属容器中，还是在有害动物很容易找到食物的纸板箱和塑料盒中？当我们没有很好地打理我们的家，而让大鼠和小鼠轻而易举地得到食物时，它们就会猖獗起来。现在检查一下你的住所周围，并确保不给有害动物任何机会在你身边安营扎寨。写份简要的程序清单，可以用来减少老鼠和你做邻居的风险。

TEACHER AND PARENT GUIDE

学科知识
Academic Knowledge

生物学	大鼠起源于欧洲和亚洲；林鼠来自北美洲；在英文中，牡（buck）指雄鼠（公鼠），牝（doe）指未交配的雌鼠，dam指怀孕的雌鼠或鼠妈妈，kitten指幼鼠；大鼠携带动物性病原体，动物性鼠疫包括口蹄疫。
化 学	鼠药的活性成分是大隆、溴敌隆、噻鼠酮和鼠得克；信息化学物质。
物 理	人类能听到频率为20~20000赫兹的声音，其中最佳听觉效果在1000~5000赫兹，这是人类语言的音频范围；与年龄相关的听觉丧失被称为老年聋；年龄超过18岁的人不能听到17400赫兹的声调；大象能听到和发出远低于人类听觉范围的声音，低至14~16赫兹，这些低频声音能让大象在相距超过1.6千米时保持联系。
工程学	训练非洲巨鼠通过气味来探测地雷和结核病；大鼠有能力通过下水管道游到厕所，这就需要新的工程形式来防止它们入内。
经济学	在经济上，大鼠作为测试毒性的介质的重要性；由大鼠和小鼠的危害造成的损失；对专业消灭有害动物的需求的增长。
伦理学	大鼠被用作实验室试验动物，而许多这类试验可以在不伤害或不杀死动物的情况下进行。
历 史	中世纪爆发的黑死病是由印鼠客蚤所携带的微生物鼠疫耶尔森菌所致；1895年，位于美国马萨诸塞州伍斯特市的克拉克大学建立了第一个用于实验室试验的家养白化褐家鼠种群；在披头士乐队的歌曲《生命中的一天》最后一个和弦之后，有一个音频为1.5万赫兹的啸声，用于让忠诚的甲壳虫乐队粉丝听到。
地 理	南极洲是唯一没有大鼠分布的地区。
数 学	小鼠种群的指数增长：每窝产仔4~16只，每年产仔7~8窝，每只小鼠每天吃3克食物、喝3毫升水、拉粪便80粒，体重12~45克；大鼠的指数增长：每窝产仔5~10只，每年产仔3~6窝，每只大鼠每天吃15~30克食物、喝15~60毫升水，体重150~300克。
生活方式	我们周围充斥着浪费，导致大鼠和小鼠就在我们身边猖獗。
社会学	在一些文化中把大鼠当作食物属于禁忌，而其他文化中却把大鼠当作美味，如印度、越南和中国台湾；一则尼日利亚的寓言故事解释了猫为什么追逐大鼠。
心理学	大鼠的心理看上去似乎与人类有些相似，特别是在创造性、进取性和适应性方面；我们一方面把很多问题的责任推卸给大鼠，另一方面又在维持大鼠种群；大鼠帮助创建了如何获得食物的心理地图；刺激-反应理论。
系统论	所有海鸟和爬行动物的灭绝中有40%~60%是由大鼠导致，而其中90%的岛屿物种的灭绝由大鼠造成。

教师与家长指南

情感智慧
Emotional Intelligence

大鼠甲

大鼠甲分享了它的喜悦，因为一项法律禁止使用鼠药。它描述了人们试图杀死有害动物的各种方法。它愿意谈论可能发生的戏剧性变化，包括一些它以前闻所未闻、见所未见的令人惊讶的新方法和新手段。大鼠甲很惊讶，但说话不失自信，它希望知道它的朋友在这方面有怎样的经历。大鼠甲指出人们所有的努力付出都效率低下，因为大鼠依然在人们身边大量出没。它毫不吝啬地夸奖人们正变得更聪明，会用辣椒了，即便采用的人还很少。大鼠甲对它的同伴表示同情。它为人类提供了一个惊喜，就是通过大鼠厌恶的气味来进行大鼠防治的方法。它很谦恭，想要向狐猴学习，与人类异地而居。

大鼠乙

大鼠乙绘声绘色地描述从胃穿孔到"电刑室"的各种危险。它也不怕冒险，并转述自己如何机智地战胜企图电死他的人们的经历。它注意到，当它毫发无伤就得到食物，电子灭鼠器却电池耗尽也一无所获，不觉言辞尖刻。大鼠乙把起因和结果搞得很清楚，它指出，如果人们想要防治鼠害，就必须停止浪费食物。大鼠乙自嘲地描述它自己误打误撞的故事和笑话，但最终都有一个快乐的结局。它热衷于持续不断地学习新事物，并与他人交流经验。

艺术
The Arts

班克斯是一位涂鸦艺术家，他把黑色幽默带到全世界的街道、墙壁和桥梁。老鼠是他最喜爱的动物之一。他采用了模版印制工艺，借助经过艺术创作制成的镂空模板，把颜料涂在一个物体上，从而能产生图像。这种工艺的优点是模板可以反复使用。因此，让我们学习如何制作一个模板：为了方便地复制你创建的图像或信息，需要在模板上把这些地方镂空。当然，老鼠应该是此次艺术活动的主题。

TEACHER AND PARENT GUIDE

思维拓展
Systems: Making the Connections

人类在有害动物防治方面非常具有创造力。为了杀死老鼠，人们无所不用其极。然而，随着对人类健康和动物权利的日益关注，用毒药让老鼠（和年幼的人类小孩）胃穿孔、用诱捕器粉碎大鼠头骨或切断四肢那样比较残忍的方式变得不受欢迎，甚至有些技术被法律明令禁止。有害动物的防治机制，现在正向创造不利于这类有害动物的环境、迫使它们寻找其他地点生存的方向转变。有些方法是纯天然的，比如使用辣椒或者利用大鼠对伤害它们耳膜的特定声音的敏感性。然而，看上去人们并没有准备好去解决导致鼠害的根本原因：过度浪费以及对资源的不明智利用。人们与其把时间和精力花在设计防治技术和方法上，还不如重新设计他们自己的生活方式以及饮食和卫生习惯，这是一种更有效地利用资源的方式。我们可以从大鼠的结论中得到启示，它指出我们应该学习其他生物，就像狐猴那样控制领地和击退不受欢迎的外来者。这里有一个例子，就是把5种让目标动物感到不舒服的臭味混合制成"鸡尾酒式的气味"，从而成功地将臭虫从床垫中驱赶出来，这个方法并不依赖有讨厌的副作用的化学品。与其发明有害的技术，不如多看看和多倾听。这是相互学习为我们提供的机遇，也是我们能以谦卑的方式来应对所有挑战的方式——在确保我们能找到更好的方式来过上幸福和健康生活的同时，也要考虑到所有与我们分享这个星球的其他生命。

动手能力
Capacity to Implement

用气味作为防治剂有助于引导行为。不同的气味将引发不同的行为。研究一下靠散播气味来开展的商业模式。你不仅要研究用臭味驱逐老鼠，还要列出那些让人们感到舒服、能舒缓压力、增强心智和决心的香味。你至少要确定十几项这样的业务，并验证是否有一项业务你愿意进行开发。

教师与家长指南

故事灵感来自

雷吉内·格里斯
Regine Gries

雷吉内·格里斯出生于德国，与她的丈夫格哈德·格里斯一起长年从事有害生物防治机制的研究工作。她是位于加拿大温哥华的西蒙弗雷泽大学格里斯实验室的助理研究员，也是对动物通讯生态学有浓厚兴趣的生物学家。人们所熟知的是，为了献身科学，她曾经让20万只臭虫对她进行叮咬。她与一个科学家团队一起，已经成功地设计出一种外激素"鸡尾酒"（混合物），能有效地将臭虫从床垫中引诱出来并驱离。格里斯实验室的团队目前正专注于信息化学物质和创新性诱饵技术的识别和筛选，这一研究将会以一种环境友好和尊重动物权利的方式控制大鼠的行为。

图书在版编目（CIP）数据

大鼠害怕香味剂：汉英对照/（比）冈特·鲍利著；（哥伦）凯瑟琳娜·巴赫绘；唐继荣译．— 上海：学林出版社，2017.10
（冈特生态童书．第四辑）
ISBN 978-7-5486-1246-9

Ⅰ．①大… Ⅱ．①冈… ②凯… ③唐… Ⅲ．①生态环境－环境保护－儿童读物－汉、英 Ⅳ．① X171.1-49

中国版本图书馆 CIP 数据核字（2017）第 143484 号

© 2017 Gunter Pauli
著作权合同登记号　图字 09-2017-532 号

冈特生态童书
大鼠害怕香味剂

作　　者——	冈特·鲍利
译　　者——	唐继荣
策　　划——	匡志强
责任编辑——	李晓梅
装帧设计——	魏来
出　　版——	上海世纪出版股份有限公司　学林出版社
	地　址：上海钦州南路 81 号　电话/传真：021-64515005
	网　址：www.xuelinpress.com
发　　行——	上海世纪出版股份有限公司发行中心
	（上海福建中路 193 号　网址：www.ewen.co）
印　　刷——	上海丽佳制版印刷有限公司
开　　本——	710×1020　1/16
印　　张——	2
字　　数——	5 万
版　　次——	2017 年 10 月第 1 版
	2017 年 10 月第 1 次印刷
书　　号——	ISBN 978-7-5486-1246-9/G.472
定　　价——	10.00 元

（如发生印刷、装订质量问题，读者可向工厂调换）